WHAT IS STEAM?

# THE ENGINEERING IN STEAM

BY THERESA EMMINIZER

**Please visit our website, www.garethstevens.com. For a free color catalog of all our high-quality books, call toll free 1-800-542-2595 or fax 1-877-542-2596.**

**Library of Congress Cataloging-in-Publication Data**
Names: Emminizer, Theresa, author.
Title: The engineering in STEAM / Theresa Emminizer.
Description: Buffalo : Gareth Stevens Publishing, [2024] | Series: What is STEAM? | Includes index.
Identifiers: LCCN 2022053337 (print) | LCCN 2022053338 (ebook) | ISBN 9781538285480 (library binding) | ISBN 9781538285473 (paperback) | ISBN 9781538285497 (ebook)
Subjects: LCSH: Engineering–Juvenile literature.
Classification: LCC TA149 .E46 2023 (print) | LCC TA149 (ebook) | DDC 620–dc23/eng/20221110
LC record available at https://lccn.loc.gov/2022053337
LC ebook record available at https://lccn.loc.gov/2022053338

Published in 2024 by
**Gareth Stevens Publishing**
2544 Clinton Street
Buffalo, NY 14224



Designer: Leslie Taylor
Editor: Theresa Emminizer

Photo credits: Cover Zivica Kerkez/Shutterstock.com; Series Art (background objects) N.Savranska/Shutterstock.com; p. 5 karelnoppe/Shutterstock.com; p. 7 Gorodenkoff/Shutterstock.com; p. 9 imtmphoto/Shutterstock.com; p. 11 Ground Picture/Shutterstock.com; p. 13 Brocreative/Shutterstock.com; p. 15 Ana Hollan/Shutterstock.com; p. 17 Chaosamran_Studio/Shutterstock.com; p. 19 Olena Yakobchuk/Shutterstock.com; p. 21 Ground Picture/Shutterstock.com.



Printed in the United States of America



CPSIA compliance information: Batch #CSGS24: For further information contact Gareth Stevens at 1-800-542-2595.

# CONTENTS



**Boldface** words appear in the Glossary.

## About STEAM

STEAM stands for science, technology, engineering, art, and math. These fields might seem very different, but they're all ways of **exploring** the world around us and finding answers to questions. In this book, you'll learn about the engineering in STEAM.

# What Is Engineering?

Engineering is the use of science and math to build better objects, or things. People who practice engineering are called engineers. Engineers plan and build machines, **systems**, and **structures**. Engineering work can be grouped into different branches: civil, electrical, mechanical, and chemical.

## Civil Engineering

Civil engineers plan and build infrastructure. Infrastructure is the systems and structures that a community needs to work. Roads, bridges, buildings, and tunnels are all parts of infrastructure. Civil engineers work in offices and on construction, or building, sites.

## Electrical Engineering

Electrical engineers study and **design** electrical systems and devices, or tools. Many of these devices are things we use every day! Batteries, smartphones, computers, and smart watches were all created, or made, by electrical engineers. Electrical engineers also design **robotics** and **drones**.

## Mechanical Engineering

Mechanical engineers work with machines and mechanical systems. A lot of mechanical engineering has to do with the parts of a moving system. A mechanical engineer might work with cars, airplanes, or roller coasters. Or they might work with **manufacturing** tools, such as sewing machines.

## Chemical Engineering

Chemical engineers work with chemicals. A chemical is matter that can be mixed with other matter to cause changes. If you've ever tie-dyed a shirt, that's a chemical reaction, or change! Chemical engineers work with food, drinks, makeup, and more.

## Engineering Skills

No matter what branch they practice, all engineers use a lot of math and science. Engineers need to be curious, ask questions, and think outside the box to come up with creative ways to solve, or fix, problems.

## Are You an Engineer?

Do you wonder about how things work? Are math and science your favorite classes? Maybe you like mechanical toys, like electronic cars or drones. Maybe you have fun **experimenting** with chemicals. You might have what it takes to be an engineer!

## Many Branches, One Goal

There are many kinds of engineers. But an engineer's **goal** is to solve problems and create things that make people's lives better. Engineers make the infrastructure, devices, machines, and foods that we use every day. Is engineering the path for you?

# GLOSSARY

**design:** To create the pattern or shape of something.

**drone:** Aircraft or flying machine that doesn't have a pilot or person to fly it.

**experiment:** To carry out scientific tests or actions to learn about something.

**explore:** To search in order to find out new things.

**goal:** Something important someone wants to do.

**manufacturing:** Making goods with machines in factories.

**robotics:** Having to do with robots.

**structure:** Something built.

**system:** A group of parts that work together.

# FOR MORE INFORMATION

## BOOKS

McAneney, Caitie. *20 Fun Facts About Famous Bridges*. New York, NY: Gareth Stevens Publishing, 2020.

Salt, Zelda. *Be an Aerospace Engineer*. New York, NY: Gareth Stevens Publishing, 2019.

## WEBSITES

**Black Girls Code**
*wearebgc.org/*
Learn more about engineering programs and how you can get involved.

**Engineer Girl**
*www.engineergirl.org/6076/Mechanical-Engineer*
Find out if you have what it takes to become an engineer!

**Publisher's note to educators and parents:** Our editors have carefully reviewed these websites to ensure that they are suitable for students. Many websites change frequently, however, and we cannot guarantee that a site's future contents will continue to meet our high standards of quality and educational value. Be advised that students should be closely supervised whenever they access the internet.

# INDEX